PERENNIAL MATH

CONTEST PROBLEMS

2011-2014

Dr. Sylvia Dean has been a gifted specialist for 25 years in Huntsville, Alabama. Dr. Dean created Perennial Math in 2011 upon completion of her doctoral degree. She has hosted many robotics, technology, and mathematics competitions throughout the years and has tirelessly worked on providing educational opportunities to gifted students around the world. It has always been her ambition to leave students with the feeling that "math is cool."

These problems are challenging and students should know up front that scoring a 3 out of 6 on any of these tests is extraordinary and above average. Great problem-solvers learn persistence and appreciation for challenge. Good luck as you encourage your students to take the quest to become great mathletes.

ADVANCED LEVEL
Grades 7-8

PERENNIAL MATH

TEST # 1

November 2011 Advanced Level

1) A computer program uses a 2 digit security key. If the sum of the digits must be a non-negative multiple of ten for the key to be valid, how many keys are possible?	
2) What is the sum of all the prime numbers between 20 and 60?	
3) In the figure shown the circle has a radius of 10 units. What is the area of the square who's corners are all points on the circle? 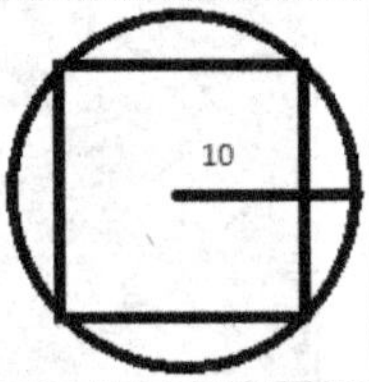	
4) Together Cathy and Felix have 60 marbles. Cathy gives Felix half her marbles and then 12 more marbles. Felix now has three times as many marbles as Cathy has. How many marbles did Felix have originally?	
5) Square tiles with 9-inch sides are used to cover a rectangular room. There is one row of green border tiles and the interior tiles are yellow. The room measures 24 feet by 18 feet. How many tiles are green?	

PERENNIAL MATH

TEST # 1

November 2011 Advanced Level Answers

1) A computer program uses a 2 digit security key, if the sum of the digits must be a non negative multiple of ten for the key to be valid, how many keys are possible?	**10**
2) What is the sum of all the prime numbers between 20 and 60?	**363**
3) In the figure shown the circle has a radius of 10 units. What is the area of the square who's corners are all points on the circle? 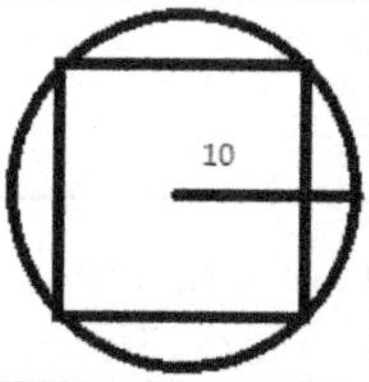	**200**
4) Together Cathy and Felix have 60 marbles. Cathy gives Felix half her marbles and then 12 more marbles. Felix now has three times as many marbles as Cathy has. How many marbles did Felix have originally?	**6**
5) Square tiles with 9-inch sides are used to cover a rectangular room. There is one row of green border tiles and the interior tiles are yellow. The room measures 24 feet by 18 feet. How many tiles are green?	**108**

PERENNIAL MATH TEST # 1

November 2011 Advanced Level Solutions

1) By enumerating the possibilities:

9+1 = 1+9 = 8+2 = 2+8 = 7+3 = 3+7 = 6+4 = 4+6 = 5+5 =10

Noting that 9+9 = 18, there can't be any keys with a digit sum of 20 or greater, so the only options are those that sum to 10, and those that sum to a number less than 10 (the only option of which is 0 = 0+0). Therefore there are **10 possible security keys**.

10

2) Add the prime numbers between 20-60 you get

23 +29 + 31 + 37 + 41 + 43 + 47 + 53 + 59 = **363**

363

3) Since r = 10 , one can make a 45-45-90 right triangles making up a quarter of the square. The sides have a length of 10 units because they are both a radius of the circle. Using the formula for area of a right triangle (Area=1/2 b x h, the area is ½ * 10 *10 , or ½ *100 = 50. Since this triangle is the area of ¼ the square, the total area is 4 * 50, or **200 square units.**

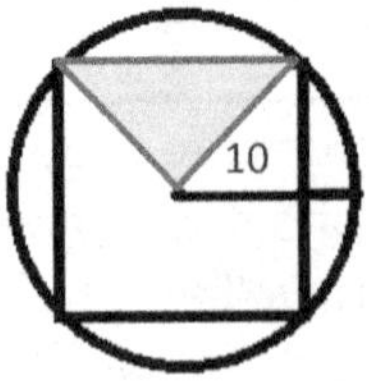

200

PERENNIAL MATH TEST # 1

November 2011 Advanced Level Solutions

4) After the exchange, Felix has 3 marbles for every 1 that Cathy has, a ratio of 3:1. That is of the 60 marbles, Felix has 3 of every 4 and Cathy has 1 or every 4. Therefore, Felix ends with 45 and Cathy with 15 marbles.

	Total	Cathy	Felix
Step 3. Cathy has 1/4 of the marbles.	60	15	45
Step 2. Cathy gives Felix 12 marbles.	60	27	33
Step 1. Cathy gives Felix 1/2 her marbles.	60	54	6

Before Cathy gave 12 of her marbles to Felix, he had 45-12=33 and Cathy had 15 +12=27. Before Cathy gave half her marbles to Felix, she had 2 x 27=54 and he had 33-27=6. Felix **originally had 6 marbles.**

6

5) The length of one side of each tile is ¾ of a foot. Thus, the floor is 24 feet divided by ¾ foot = 32 tiles long and 18 divided by ¾ = 24 tiles wide. The entire floor consists of 32 x 24 = 768 tiles. The yellow portion of the floor consists of 30 x 22 = 660 tiles. Thus, 768-660 =**108 tiles that are green**.

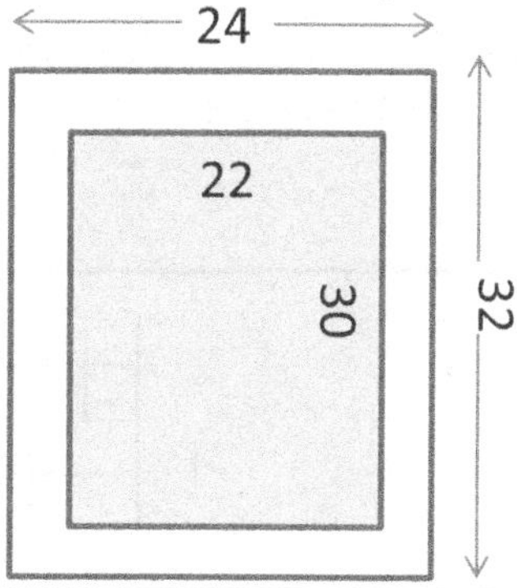

108

PERENNIAL MATH

TEST # 2

December 2011 Advanced Level

1) The product of two whole numbers is 64. The average of the two numbers is 10. What is the smallest of the two numbers?	
2) What is the mean of all the counting numbers less than 100 which are multiples of 4?	
3) There are 150 students in the 7th grade. While 75 of them play sports and 60 of them play musical instruments, 20 of them play neither sports nor instruments. How many students play sports, but not musical instruments?	
4) 3 + 7= 10 and 5 + 5 = 10 are two ways to express 10 as the sum of two prime numbers. In how many ways can 60 be expressed as the sum of two prime numbers?	
5) How many squares of any size are in this figure? 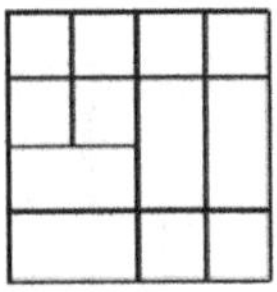	

PERENNIAL MATH

TEST # 2

December 2011 Advanced Level Answers

1) The product of two whole numbers is 64. The average of the two numbers is 10. What is the smallest of the two numbers?	**4**
2) What is the mean of all the counting numbers less than 100 which are multiples of 4?	**50**
3) There are 150 students in the 7th grade. While 75 of them play sports and 60 of them play musical instruments, 20 of them play neither sports nor instruments. How many students play sports, but not musical instruments?	**70**
4) 3 + 7= 10 and 5 + 5 = 10 are two ways to express 10 as the sum of two prime numbers. In how many ways can 60 be expressed as the sum of two prime numbers?	**6**
5) How many squares of any size are in this figure? 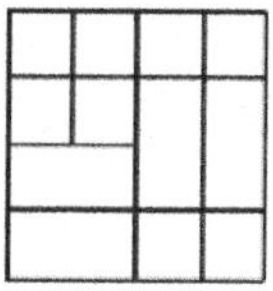	**15**

PERENNIAL MATH

TEST # 2

December 2011 Advanced Level Solutions

1) If you write the factors of 60 and add them up, then 4 + 16 = 20 and the average of 20/2 = 10. The lowest number of the pair whose product is 64 and average is 10 is **4**.

Factors		Sum	Average
1	64	65	32.5
2	32	34	17
4	16	20	10
8	8	16	8

4

2) Listing the multiples of 4 and going to 99, there would be 24 multiples. The sum of these 24 multiples is 1200. 1200 divided by 24 equals **50**.

50

3) Since 20 students play neither sports nor instruments, there are 130 students who play either sports or instruments. Because 75 + 60 = 135, then 5 students were counted twice. This is only possible if these 5 students play both sports and instruments. Therefore, 5 of the 75 students who play sports also play instruments. Then 75 - 5 = **70 students play sports, but not instruments**. See the Venn diagram.

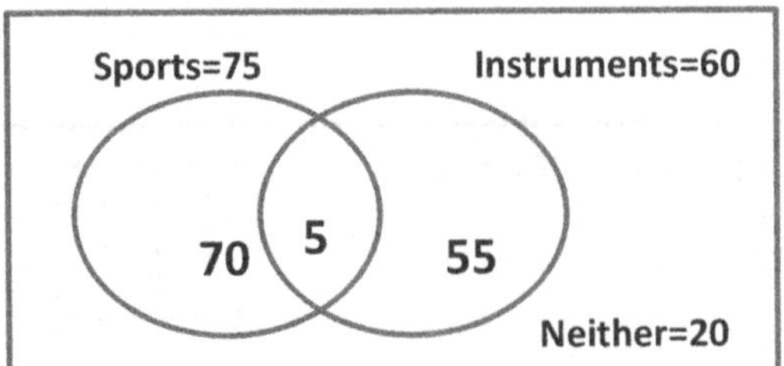

70

PERENNIAL MATH TEST # 2

December 2011 Advanced Level Solutions

4) List the prime numbers to 60: 2, 3, 5, 7, 11, 13, 17, 19, 23, 17, 29, 31, 37, 41, 43, 47, 53. Most of these can be paired up to produce a sum of 60 with the exception of 2, 3, 5, and 11.

7 + 53 = 60
13 + 47 = 60
17 + 43 = 60
19 + 41 = 60
23 + 37 = 60
29 + 41 = 60

There are 6 pairs of prime numbers that have a sum of **60**.

6

5) There are 8 small squares, 4 medium squares, 2 large squares, and 1 extra large square for a total of **15 squares**.

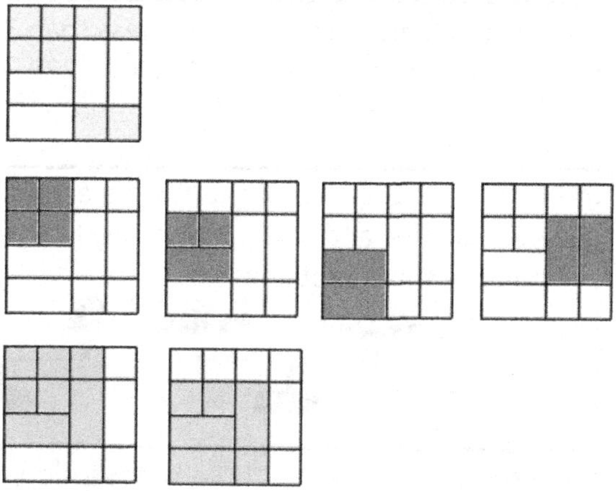

15

PERENNIAL MATH

TEST # 3

January 2012 Advanced Level

1) What is the sum of the first 20 even numbers greater than zero?	
2) Tony stacks his tennis balls and continues the triangular pattern out to the 10th stack, how many balls does Tony have altogether? 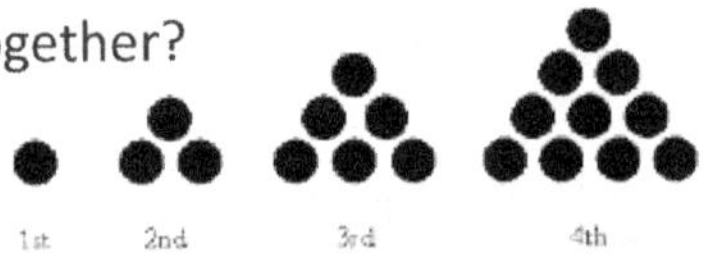	
3) If B and C are prime numbers less than 35, what is the value of C if: $(3 \times B) + (5 \times C) = 224$	
4) Using the diagrams of the two balanced scales of lizards, rabbits, and mice, how many lizards will it take to balance three rabbits? 	
5) A rectangular field is three times as long as it is wide and has an area of 1323 square yards. Fencing is needed for the perimeter and to divide the field into three square sections. How many feet of fencing are needed?	

PERENNIAL MATH

TEST # 3

January 2012 Advanced Level Answers

1) What is the sum of the first 20 even numbers greater than zero?	**420**
2) Tony stacks his tennis balls and continues the triangular pattern out to the 10th stack, how many balls does Tony have altogether? 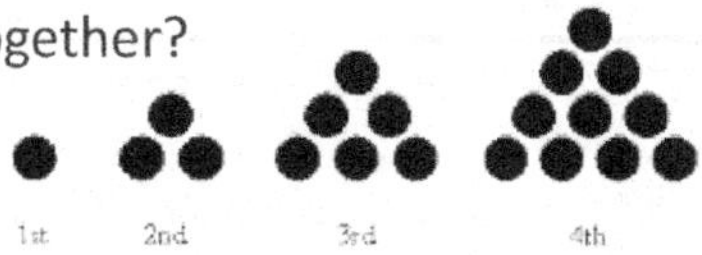	**220**
3) If B and C are prime numbers less than 35, what is the value of C if: $(3 \times B) + (5 \times C) = 224$	**31**
4) Using the diagrams of the two balanced scales of lizards, rabbits, and mice, how many lizards will it take to balance three rabbits? 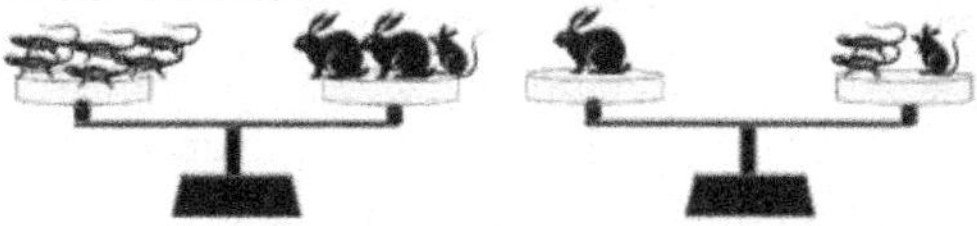	**8**
5) A rectangular field is three times as long as it is wide and has an area of 1323 square yards. Fencing is needed for the perimeter and to divide the field into three square sections. How many feet of fencing are needed?	**630**

PERENNIAL MATH

TEST # 3

January 2012 Advanced Level Solutions

1) You could write out the numbers and add them up. 2 + 4 + 6 + 8 + 10 + 12 + 14 + 16 + 18 +20 +22 + 24 + 26 + 28 + 30 + 32 + 34 + 36 + 38 + 40 = 420. Notice the first and last numbers can be paired up

2 + 4 + 6 +8 + ... + 16 +18 + 20

40 + 38 + 36 + 34 + ... + 26 +24 + 22

and there are 10 pairs of numbers whose sum is 42. Ten pairs times 42 is 420. Based on number theory of Carl F. Gauss- Solution: Let N represent the number of numbers, then N(N+1), is the sum of the first N even numbers.

20 * (20 + 1)= 20 * 21 = **420**

420

2) So the first triangular number is 1, the second is 3 (= 1 + 2), the third is 6 (= 1 + 2 + 3), the fourth is 10 (= 1 + 2 + 3 + 4) and the fifth is 15 (= 1 + 2 + 3 + 4 + 5). For each of the stacks add the number of balls for the number of stack it is to the previous stack of balls . You could make a chart to see how many balls are in each stack and add all the stacks together.

Stack	1	2	3	4	5	6	7	8	9	10
Balls	1	3	6	10	15	21	28	36	45	55
Total	1	4	10	20	35	56	84	120	165	220

220

3) If B and C are both different prime numbers less than 35, what is the value of C if (3 x B) + (5 x C) = 224. Rewrite the problem as 3B + 5C = 224.
Use the divisibility of 5 rule for 5C= 224- 3B. Since 5C must end in 0 or 5, then 224-3B must end in 4 or 9. Therefore, if you list out the prime numbers up to 35, {2, 3, 5, 7, 11, 13, 17, 19, 23, 29, 31}, there are no even numbers that are prime excluding 2- so the number would have to end in 9.
The only prime numbers that could possibly give you a number ending in 9 when multiplied by 3, are 3, 13, and 23. Try these out for B and you have 224-3(3) = 215; 224 – 3 (13)= 185; and 224- 3(23) = 155. Divide these answers by 5 to solve for C and you get 215/5= 43; 185/5=37; and 155/5=31. C must equal **31** since it has to be less than 35.
That would make B equal 23.

31

PERENNIAL MATH

TEST # 3

January 2012 Advanced Level Solutions

4) Use algebra and rewrite the problem as 6L=2R + M and R = 2L + M.

$$6L + (2L + M) = 2R + M + (R)$$

$$8L + M - M = 3R + M - M$$

$$8L = 3R$$

Using the left scale because it already has 2 rabbits, If you add a rabbit R to each side and substitute two lizards and a mouse on the left side you will have and you can subtract a mouse from each side leaving you with **8 lizards** weighing the same as 3 rabbits.

8

5) Given that Area = L x W = 1323 yd^2 and L = 3W.
Substituting 3W for L, 3W x 1W = 3 W^2 = 1323 yd^2
Dividing each side by 3, W^2 = 441 yd^2
Taking the square root of each side, W = 21 yd so L = 3 x 21 yd = 63 yd
Perimeter = 2 x (L + W) = 2 x (63 yd + 21 yd) = 2 x 84 yd = 168 yd. You must also add two more sides of 21 in order to account for the fencing to make the field into 3 equal squares. So 168yd + 42yd= 210yd. Now convert that into feet 210 x 3= **630 ft.**

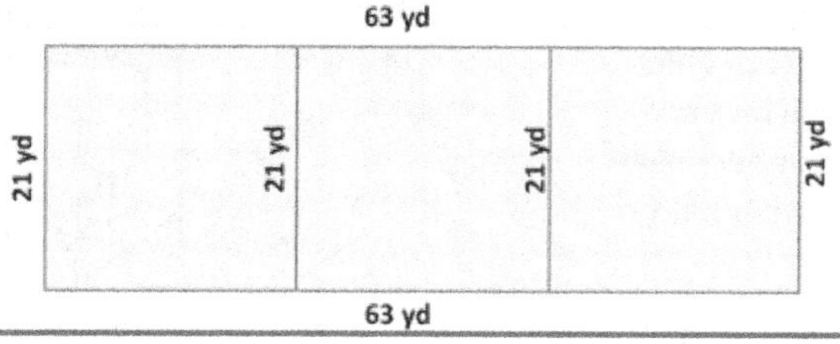

630

PERENNIAL MATH

TEST # 4

February 2012 Advanced Level

1) Three pumpkins and two melons weigh 32 pounds. Four pumpkins and three melons weigh 44 pounds. All pumpkins weigh the same and all melons weigh the same. How many pounds does two pumpkins and one melon weigh?	
2) Mrs. Smith recorded the test grades for all 25 students in her seventh grade class. She used the grades recorded to calculate the class average to be 72. Jamia's grade of 86 was incorrectly marked as 36. What was the correct class average for the test?	
3) Sandy received birthday money from her grandmother. She put 2/3 of the money in her bank account, spent $25 on a sweater, and spent half the remaining money on a necklace. If Sandy had $15 left over, how much money did she receive from her grandmother?	
4) A bat ate a total of 1050 mosquitos on four consecutive nights. Each night it ate 25 more than on the night before. How many did it eat on the third night?	
5) If this shape were glued together and fell into a bucket of paint, how many cubes would have paint on only 3 sides? 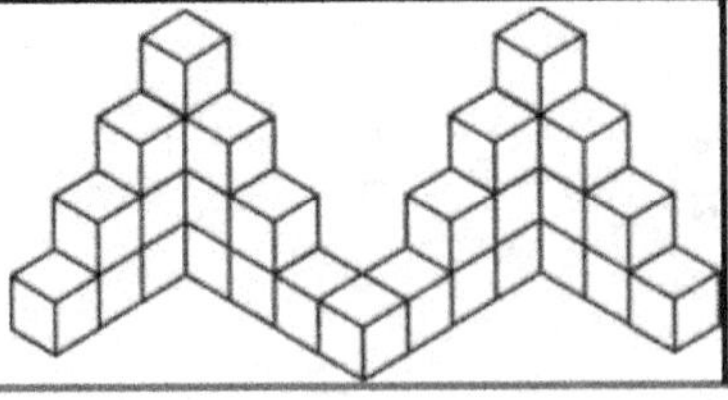	

PERENNIAL MATH

TEST # 4

February 2012 Advanced Level Answers

1) Three pumpkins and two melons weigh 32 pounds. Four pumpkins and three melons weigh 44 pounds. All pumpkins weigh the same and all melons weigh the same. How many pounds does two pumpkins and one melon weigh?	**20**
2) Mrs. Smith recorded the test grades for all 25 students in her seventh grade class. She used the grades recorded to calculate the class average to be 72. Jamia's grade of 86 was incorrectly marked as 36. What was the correct class average for the test?	**74**
3) Sandy received birthday money from her grandmother. She put 2/3 of the money in her bank account, spent $25 on a sweater, and spent half the remaining money on a necklace. If Sandy had $15 left over, how much money did she receive from her grandmother?	**165**
4) A bat ate a total of 1050 mosquitos on four consecutive nights. Each night it ate 25 more than on the night before. How many did it eat on the third night?	**275**
5) If this shape were glued together and fell into a bucket of paint, how many cubes would have paint on only 3 sides? 	**10**

PERENNIAL MATH

TEST # 4

February 2012 Advanced Level Solutions

1) If you subtract the first scenario from the second scenario you will discover that 1 pumpkin and 1 melon weigh 12 lbs. Now you can deduce that 2 pumpkins and 2 melons weigh 24 lbs. Subtract that from the first statement and discover that pumpkins are 8 lbs. If pumpkins are 8, then the melon weighs 4 because P + M = 12

$$\begin{array}{r} 4P + 3M = 44 \\ - \underline{3P + 2M = 32} \\ P + M = 12 \end{array} \qquad \begin{array}{r} 3P + 2M = 32 \\ - \underline{2P + 2M = 24} \\ P \quad\quad = 8 \end{array}$$

2 pumpkins (8 lbs.) and 1 melon (4 lbs.) weigh **20** lbs.

20

2) If the class average equals the sum of all the grades divided by the number of grades (students) then the sum of the grades can be calculated as 25 x 72 or 1800. Since the 86 was recorded as 36, fifty points must be added to the total sum. 1850 divided by 25 would yield a new class average of 74.

x / 25 = 72

x = 1800 = the sum of all the grades

1800 - 36 + 86 = 1850

1800 / 25 = **74%**

74

3) There are various ways to solve this problem. One way is to draw a diagram to help separate out the events. 2/3 of the money was put in the bank. The money that remained is 1/3. Of that third, $25 was spent on a sweater and then half of the remainder on a necklace. $15 remained and it was half of the remainder of money so the necklace had to cost $15. A third of the money would be $25 + $15 + $15 or $55. The total amount of money given to her is 3 x $55 or **$165**.

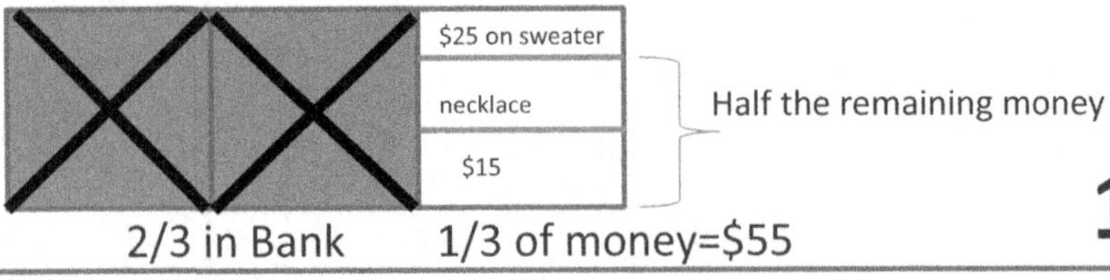

165

PERENNIAL MATH

TEST # 4

February 2012 Advanced Level Solutions

4) X + (X + 25) + (X + 50) + (X + 75) =1050

4X + 150 = 1050

4X = 900

X = 225

The third night would be 225 + 50 = **275**

275

5) There are 33 total blocks in this shape. The blocks that are highlighted have only 3 sides that have paint on them because they have a block touching on two sides and the top leaving 3 sides untouched. There are 3 hidden blocks in the top layers of each sides of the shape. Each of those layers will have the bottom block that is in the corner with paint on three sides because that bottom block is only touched on two sides and the top. So the 8 that are clearly seen and the two hidden blocks add up to **10 blocks** who have paint only on three sides.

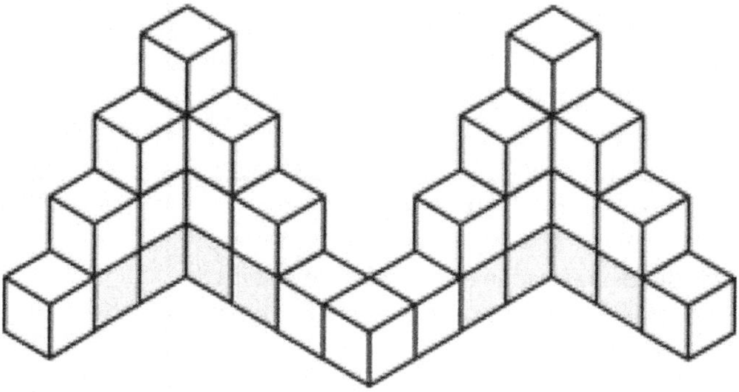

10

PERENNIAL MATH

TEST # 5

March 2012 Advanced Level

1) You are given a set containing the first twenty-five counting numbers {1, 2, 3, . . . 23, 24, 25}. What is the difference between the sum of the numbers in this set that are divisible by 2, 3, or 4 and the sum of the numbers in this set that are NOT divisible by 2, 3, or 4?	
2) A 5-digit even number is greater than 20,000 but less than 70,000 and no digit is repeated. The digit in the hundreds place is 3 times the digit in the ten thousands place. The ones digit is a prime number and is less than the ten thousands digit . The thousands digit is half the sum of the ones, hundreds, and ten thousands digits. Lastly, the tens digit is the product of two other digits. What is the sum of all the digits in this 5-digit number?	
3) At a senior class party, one-half the class drank only punch, one-third of the class drank only cola, fifteen people drank neither, and nobody drank both. How many more students drank punch than cola at the party?	
4) How many total triangles of any size do you see in this diagram? 	
5) A dark pantry contains 8 cans of black beans, 9 cans of green beans, 13 cans of baked beans, 17 cans of lima beans, and 18 cans of pinto beans. How many cans must be selected from the pantry to guarantee that 12 cans of the same type of bean are chosen?	

PERENNIAL MATH

TEST # 5

March 2012 Advanced Level Answers

1) You are given a set containing the first twenty-five counting numbers {1, 2, 3, . . . 23, 24, 25}. What is the difference between the sum of the numbers in this set that are divisible by 2, 3, or 4 and the sum of the numbers in this set that are NOT divisible by 2, 3, or 4?	**83**
2) A 5-digit even number is greater than 20,000 but less than 70,000 and no digit is repeated. The digit in the hundreds place is 3 times the digit in the ten thousands place. The ones digit is a prime number and is less than the ten thousands digit . The thousands digit is half the sum of the ones, hundreds, and ten thousands digits. Lastly, the tens digit is the product of two other digits. What is the sum of all the digits in this 5-digit number?	**27**
3) At a senior class party, one-half the class drank only punch, one-third of the class drank only cola, fifteen people drank neither, and nobody drank both. How many more students drank punch than cola at the party?	**15**
4) How many total triangles of any size do you see in this diagram? 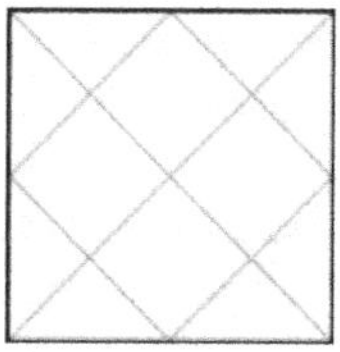	**20**
5) A dark pantry contains 8 cans of black beans, 9 cans of green beans, 13 cans of baked beans, 17 cans of lima beans, and 18 cans of pinto beans. How many cans must be selected from the pantry to guarantee that 12 cans of the same type of bean are chosen?	**51**

PERENNIAL MATH

TEST # 5

March 2012 Advanced Level Solutions

1) The numbers in the set that are divisible by 2, 3 or 4 are

{2, 3, 4, 6, 8, 9, 10, 12, 14, 15, 16, 18, 20, 21, 22, 24}.

The sum of these numbers is 204

The numbers in the set that are not divisible by 2, 3 or 4 are

{1, 5, 7, 11, 13, 17, 19, 23, 25}.

The sum of these numbers is 121

And 204 - 121 = **83**

83

2) Make a checklist of the requirements for the number:

- ✓ It is even
- ✓ Between 20,000 and 70,000
- ✓ No digits repeated
- ✓ The digit in the hundreds place is 3 times the digit in the ten thousands place.
- ✓ The ones digit is a prime number and is less than the ten thousands digit .
- ✓ The thousands digit is half the sum of the ones, hundreds, and ten thousands digits.
- ✓ The tens digit is the product of two other digits.

Assign letters for each digit ABCDE. Since the number is even and the ones digit is a prime, the ones digit E must be 2. C is equal to 3A and A must be between 2 and 7. A cannot be 2 because 2 is already used, so it must be 3 because any other digit would make C a two digit number. A is 3 and C equals 9. Since B is equal to (2 + 9 + 3)/2, B = 7. D must be 6 because it is the product of two other digits and the only 2 digits that can yield a one-digit product is 2 and 3. The number is 37,962. The sum of its digits is 3 + 7 + 9 + 6 + 2 = **27**

27

3) As 1/2 + 1/3 = 3/6 + 2/6 = 5/6, then we know that 1/6 drank neither.
Since 1/6 drank neither and we know that 15 drank neither then there must have been 6 x 15 = 90 people at the class party. Since half of those drank punch and ½ is equal to 3/6, then half of the students =45. One third of the students is equal to 2/6 or 30 students who drank cola. So the number of students who drank punch minus the number of students who drank cola is 45-30 or 15. There were **15 more students** who drank punch than cola.

15

PERENNIAL MATH

TEST # 5

March 2012 Advanced Level Solutions

4) If you count the different sizes of triangles and add them up you have:

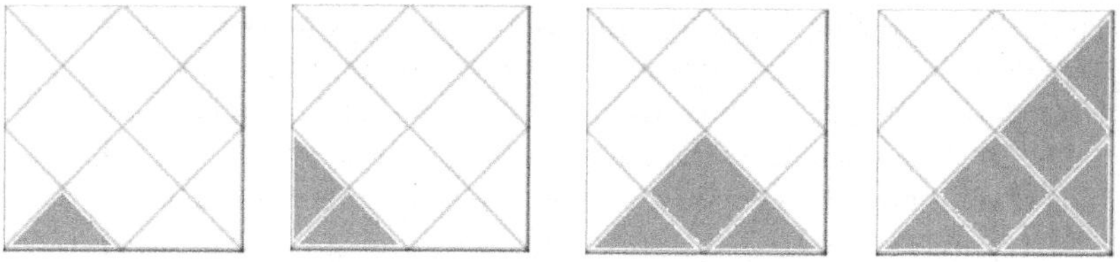

8 of these | 4 of these | 4 of these | 4 of these

So there are 8 + 4 + 4 + 4 = **20 triangles** in the diagram.

20

5) You have to eliminate all the cans that could be selected and since there are fewer than 12 cans of both the black beans and green beans all of those cans must be selected first, so 17 cans so far. Then, there are more than 12 cans of each of the three remaining bean types so 11 of each could be selected giving 17 + 3(11) = 50 cans. Then the next can that would be selected would definitely be one of the three remaining bean types (baked, lima , or pinto) resulting in a total of **51 cans of beans**.

51

PERENNIAL MATH

TEST #1

November 2012 Advanced Level

1) In a guessing game, 5 people made guesses. Zack was four under, Mary was three under, Debbie was two over, Timothy was one over, and Suzanne was four over. The numbers were 34, 31, 32, 27, and 26. What was the correct number?

2) Mr. Tate and Mr. Watkins brought some pumpkins to the farmers market. When Mr. Tate put 3 pumpkins and Mr. Watkins put 4 pumpkins on the scale it weighed 37 pounds. When Mr. Watkins put 3 pumpkins and Mr. Tate put 4 pumpkins on the scale it weighted 40 pounds. Assuming that his pumpkins weigh the same, how many pounds did one of Mr. Tate's pumpkins weigh?

3) In the figure shown, all the shapes represent different numbers and the total sum of each row and column is shown. What is the sum of the top row?

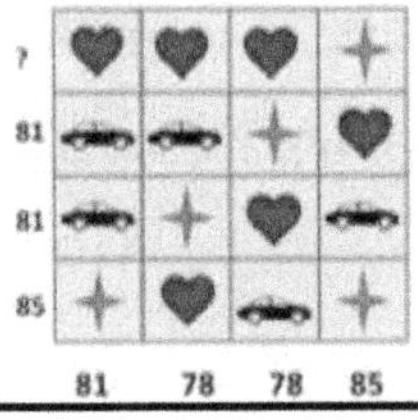

4) On the planet Benxoip, there are some 2-headed animals with 3 legs and some 2-headed animals with 2 legs. When Morcuvius counts, he sees there are 106 heads and136 legs in all. How many 3-legged animals are there?

5) Bonnie's inventory of her flower shop sales during the month of February reflects that the ratio of roses to lilies sold was 6:1. If there were a total of 749 lilies sold, how many flowers did Bonnie sell in February?

6) Kevin went to Burger Champs and spent half the money he had on lunch. Then he spent $1.75 on dessert. He had $4.25 left in his wallet. How many dollars did he have when he arrived at Burger Champs?

PERENNIAL MATH

TEST #1

November 2012 Advanced Level Answers

1) In a guessing game, 5 people made guesses. Zack was four under, Mary was three under, Debbie was two over, Timothy was one over, and Suzanne was four over. The numbers were 34, 31, 32, 27, and 26. What was the correct number?	**30**
2) Mr. Tate and Mr. Watkins brought some pumpkins to the farmers market. When Mr. Tate put 3 pumpkins and Mr. Watkins put 4 pumpkins on the scale it weighed 37 pounds. When Mr. Watkins put 3 pumpkins and Mr. Tate put 4 pumpkins on the scale it weighted 40 pounds. Assuming that his pumpkins weigh the same, how many pounds did one of Mr. Tate's pumpkins weigh?	**7**
3) In the figure shown, all the shapes represent different numbers and the total sum of each row and column is shown. What is the sum of the top row? 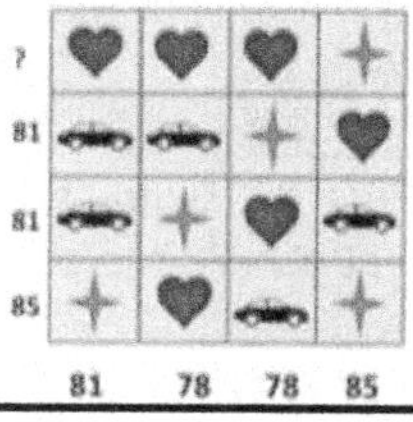	**75**
4) On the planet Benxoip, there are some 2-headed animals with 3 legs and some 2-headed animals with 2 legs. When Morcuvius counts, he sees there are 106 heads and136 legs in all. How many 3-legged animals are there?	**30**
5) Bonnie's inventory of her flower shop sales during the month of February reflects that the ratio of roses to lilies sold was 6:1. If there were a total of 749 lilies sold, how many flowers did Bonnie sell in February?	**5243**
6) Kevin went to Burger Champs and spent half the money he had on lunch. Then he spent \$1.75 on dessert. He had \$4.25 left in his wallet. How many dollars did he have when he arrived at Burger Champs?	**12**

PERENNIAL MATH TEST #1

November 2012 Advanced Level Solutions

1) In a guessing game, 5 people made guesses. Zack was four under, Mary was three under, Debbie was two over, Timothy was one over, and Suzanne was four over. The numbers were 34, 31, 32, 27, and 26.
Since the numbers are evened out with (four under and three under) = (two over, one over, and four over), then an average will give the number. The numbers 34, 31, 32, 27, and 26 added together =150. Divide 150 by 5 and the **average is 30**. 30 is the number, and 26 is 4 under, 27 is 3 under, 32 is 2 over, 31 is 1 over, and 34 is 4 over.

30

2) Mr. Tate put 3 pumpkins and Mr. Watkins put 4 pumpkins on the scale it weighed 37 pounds. Mr. Watkins put 3 pumpkins and Mr. Tate put 4 pumpkins on the scale it weighted 40 pounds. Use algebra. Let X =Mr. Tate's pumpkin and Y = Mr. Watkins' pumpkin.

$$\begin{array}{l} 3X + 4Y = 37 \\ \underline{4X + 3Y = 40} \\ 7X + 7Y = 77 \text{ (Divide both sides by 7 and } X + Y = 11) \end{array}$$

Substitute X + Y = 11 back in the equation to solve for X or Mr. Tate's pumpkin.

$$\begin{array}{l} 4X + 3Y = 40 \\ \underline{-3X + 3Y = 33} \\ \mathbf{X \quad = 7} \end{array}$$

7

3) Use algebra to solve for 3H + S

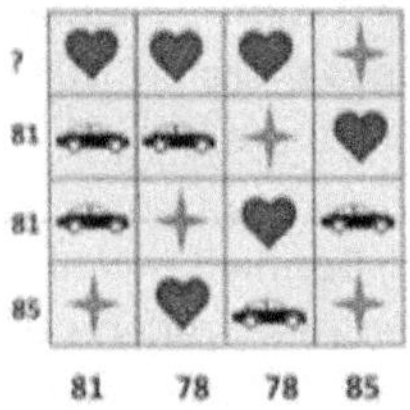

$$\begin{array}{l} 2C + S + H = 81 \\ C + 2S + H = 85 \\ \underline{C + S + 2H = 78} \\ 4C + 4S + 4H = 244 \text{ divide both sides by 4 and } C + S + H = 61. \end{array}$$

Substitute C + S + H = 61 back in the equations.

$$\begin{array}{l} C + S + 2H = 78 \\ -\ \underline{C + S + H = 61} \\ H = 17 \end{array} \qquad \begin{array}{l} C + 2S + H = 85 \\ -\ \underline{C + S + H = 61} \\ S = 24 \end{array}$$

Top row equals 3H + S or 3(17) + 24 = **75**

75

PERENNIAL MATH

TEST # 1

November 2012 Advanced Level Solutions

4) On the planet Benxoip, there are some 2-headed animals with 3 legs and some 2-headed animals with 2 legs. When Morcuvius counts, he sees there are 106 heads and136 legs in all. Use a chart to solve the problem. To calculate heads, add the animals and multiply by 2. To calculate the legs multiply the 3-legged animals by 3 and multiply the 2-legged animals by 2 and add the legs together. There are **thirty 3-legged animals**.

Total Heads	106
2 heads/3 legs	30
2 heads/2 legs	23
Total Legs	136

30

5) Bonnie's inventory of her flower shop sales during the month of February reflects that the ratio of roses to lilies sold was 6:1. If there were a total of 749 lilies sold, how many flowers did Bonnie sell in February?

Since the ratio of roses to lilies is 6:1, then there were 749 x 6= 4494 roses. The total amount of flowers for February is 4494 roses + 749 lilies = **5243 flowers**.

5243

6) Kevin went to Burger Champs and spent half the money he had on lunch. Then spent $1.75 on dessert. He had $4.25 left in his wallet. How many dollars did he have when he arrived at Burger Champs?

Work backwards: $4.25 + $1.75 = $6, $6 x 2 = **$12**

Draw a Picture and Solve

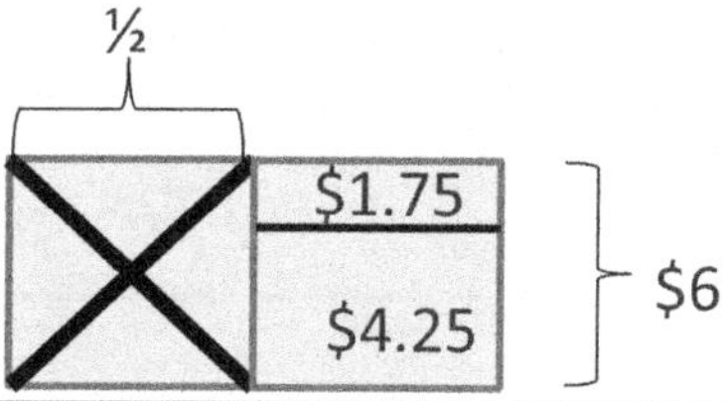

12

PERENNIAL MATH

TEST # 2

December 2012 Advanced Level

1) 4/7 of the Christmas cookies were eaten on Christmas day. The next day your brother ate 1/2 of what was left and he left you 18. How many cookies were there in the beginning?	
2) Akamu is twice as old as Kailani was three years ago. The total of their ages now is 36. How old is Kailani now?	
3) If you continue the pattern here, what letter will the number 202 come under?	
4) Clyde counted the elves and reindeer at the Christmas photo booth. He noticed there were three and a half times the number of legs as the number of heads and the total of the legs and heads was 72. How many reindeer were counted?	
5) The stadium has a parking lot that will hold 1000 vehicles. 2/5 of the parking spaces are for trucks. When Bob went to park his truck, there were 200 trucks and some cars in the parking lot. The parking lot was 3/4 full. How many cars were in it?	
6) Justin is stacking ice cubes to build a pyramid over the holiday. Justin can stack 15 ice cubes in an hour, but 2 ice cubes melt every 15 minutes. How many hours will it take him to stack 91 ice cubes?	

Table for question 3:

A	B	C	D	E	F	G	H
2	3	6	7	10	11	14	15
18	19	22	23	26	27	30	31
34	35	38	39	42	43	46	47
50	51	54	55	58	59	62	63

PERENNIAL MATH

TEST #2

December 2012 Advanced Level Answers

1) 4/7 of the Christmas cookies were eaten on Christmas day. The next day your brother ate 1/2 of what was left and he left you 18. How many cookies were there in the beginning?	**84**
2) Akamu is twice as old as Kailani was three years ago. The total of their ages now is 36. How old is Kailani now?	**14**
3) If you continue the pattern here, what letter will the number 202 come under?	**E**
4) Clyde counted the elves and reindeer at the Christmas photo booth. He noticed there were three and a half times the number of legs as the number of heads and the total of the legs and heads was 72. How many reindeer were counted?	**12**
5) The stadium has a parking lot that will hold 1000 vehicles. 2/5 of the parking spaces are for trucks. When Bob went to park his truck, there were 200 trucks and some cars in the parking lot. The parking lot was 3/4 full. How many cars were in it?	**550**
6) Justin is stacking ice cubes to build a pyramid over the holiday. Justin can stack 15 ice cubes in an hour, but 2 ice cubes melt every 15 minutes. How many hours will it take him to stack 91 ice cubes?	**13**

Table for question 3:

A	B	C	D	E	F	G	H
2	3	6	7	10	11	14	15
18	19	22	23	26	27	30	31
34	35	38	39	42	43	46	47
50	51	54	55	58	59	62	63

PERENNIAL MATH TEST #2

December 2012 Advanced Level Solutions

1) Draw a picture or use algebra. 1/7 = 12 So 7 x 12 = **84 cookies**.

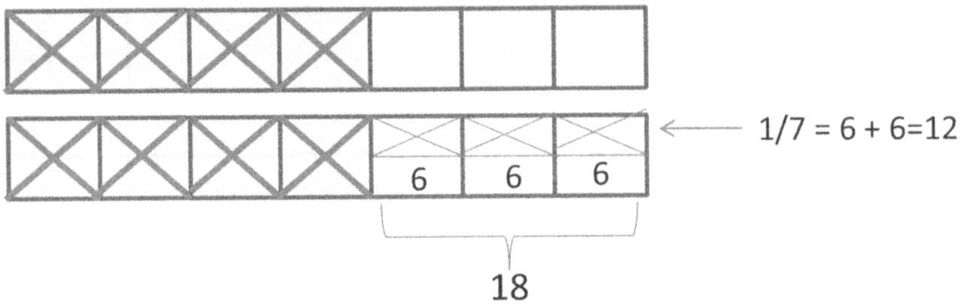

84

2) Use algebra or make a table. Akamu is twice as old as Kailani was three years ago. The total of their ages is 36. Akamu = 2 (K -3) so

2(K-3) + K =36

2K – 6 + k = 36

3K -6 = 36

3K – 6 + 6 = 36 +6

3K = 42

Akamu	Kailani	Total (36)
2 (K-3)	K	36
22	14	36

K = 42/3 OR 14 so K-3 =11 Kailani is 14 now and was 11 years old 3 years ago. Akamu is 2(K-3) or 2(11) = 22 years old. 22 + 14 = 36 years old. **Kailani is 14**

14

3) If you continue the pattern here, the only possibilities for 202 are A, C, E, and G because it is an even number. The numbers under each letter are increasing by 16. You can subtract the last number of the column from 202 and see if it is divisible by 16.

202-50= 152.

152/16 leaves a remainder. 202-54=148.

148/16 leaves a remainder and so does

202 – 62 = 140/16. The only number to not leave a remainder is 202-58=144/16=9 so **Column E is where 202 will appear**.

A	B	C	D	E	F	G	H
2	3	6	7	10	11	14	15
18	19	22	23	26	27	30	31
34	35	38	39	42	43	46	47
50	51	54	55	58	59	62	63

E

PERENNIAL MATH TEST #2

December 2012 Advanced Level Solutions

4) Use algebra or make a chart. There are 3 ½ times as many legs as heads. Let X represent heads. 3.5x + x = 72, 4.5x = 72   x= 72/4.5 = 16 so there are 16 heads. Working from 16 heads, you can use your chart to calculate legs. There are **12 Reindeer** and 4 elves.

Elves	4
Reindeer	12
Heads	16
Legs	56
Total Heads + Legs (72)	72

12

5) Draw a picture. Draw the parking lot in 5ths and then separate these into fourths. The parking lot is ¾ full which is 15/20. Each fourth holds 50 since 50 x 20 =1000. There would be 15 lots filled and 5 lots empty. There are 11 lots with trucks. 11 x 50 = **550**

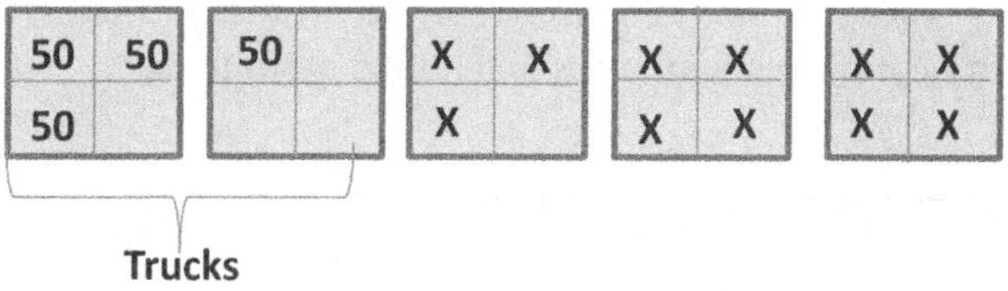

550

6) Justin stacks 15 ice cubes in an hour, but 2 ice cubes melt every 15 minutes for a total of 8 every hour. He is able to actually stack 15-8 or 7 in an hour. It will take 91 ÷ 7 = **13 hours** to stack the 91 cubes.

13

PERENNIAL MATH

TEST # 3

January 2013 Advanced Level

1) Use the order of operations to solve:

$$3 + 2 \times 4 + 2^3 - 1 =$$

2) What is the sum of all the composite numbers between 20 and 60?

3) How many rectangles of any size are In this picture?

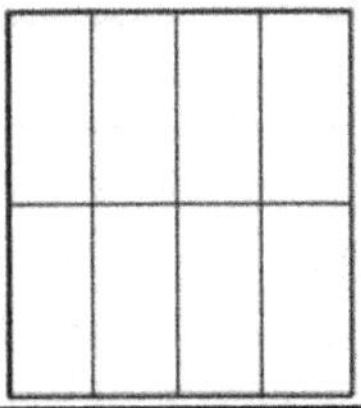

4) Joe can mow his mother's lawn in 60 minutes. His brother Ricky can mow it in 90 minutes. How many minutes will it take them to do it together?

5) Burger Max uses 2 potatoes per order of French Fries. There are 20 potatoes in a 5 pound bag which costs $2.50 per bag. How much money does the restaurant take in on a day that they use 400 pounds of potatoes if they charge $1 for an order of Fries?

6) A Class of 50 students completed a survey on what pets they like. Everyone liked at least one pet. 4 students liked Cats and Fish but not Dogs. 12 students liked Dogs and Fish but not Cats. 10 students liked Cats and Fish. 18 students liked Cats and Dogs. Altogether, 30 students liked Cats, 32 students liked Dogs, and 28 students liked Fish. How many students like fish only?

PERENNIAL MATH

TEST #3

January 2013 Advanced Level Answers

1) Use the order of operations to solve: $3 + 2 \times 4 + 2^3 - 1 =$	**18**
2) What is the sum of all the composite numbers between 20 and 60?	**1197**
3) How many rectangles of any size are In this picture? 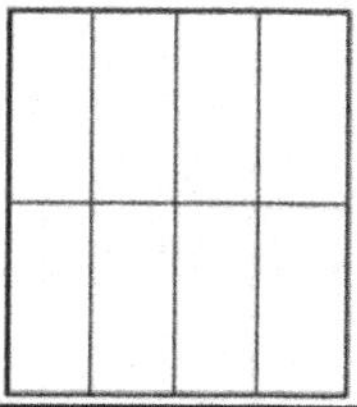	**30**
4) Joe can mow his mother's lawn in 60 minutes. His brother Ricky can mow it in 90 minutes. How many minutes will it take them to do it together?	**36**
5) Burger Max uses 2 potatoes per order of French Fries. There are 20 potatoes in a 5 pound bag which costs $2.50 per bag. How much money does the restaurant take in on a day that they use 400 pounds of potatoes if they charge $1 for an order of Fries?	**$600**
6) A Class of 50 students completed a survey on what pets they like. Everyone liked at least one pet. 4 students liked Cats and Fish but not Dogs. 12 students liked Dogs and Fish but not Cats. 10 students liked Cats and Fish. 18 students liked Cats and Dogs. Altogether, 30 students liked Cats, 32 students liked Dogs, and 28 students liked Fish. How many students like fish only?	**6**

PERENNIAL MATH TEST #3

January 2013 Advanced Level Solutions

1) PEMDAS (Parenthesis, Exponents, Multiplication, Division, Addition, Subtraction)

$3 + 2 \times 4 + 2^3 - 1 = 3 + [2\times4] + [2\times2\times2] - 1 = 3 + 8 + 8 - 1 =$ **18**

18

2) Composite numbers between 20 and 60 are 21, 22, 24, 25, 26, 27, 28, 30, 32, 33, 34, 35, 36, 38, 39, 40, 42, 44, 45, 46, 48, 49, 50, 51, 52, 54, 55, 56, 57, 58. Their sum is

1197

3) How many rectangles of any size are In this picture?

30

Types of Rectangles (Made from this # rectangles	How Many Rectangles
1	8
2	10
3	4
4	5
6	2
8	1

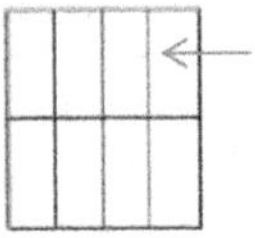

Rectangles made of 2 rectangles = 10

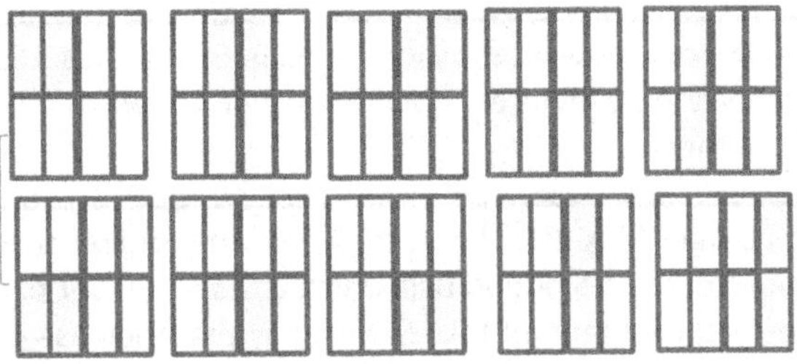

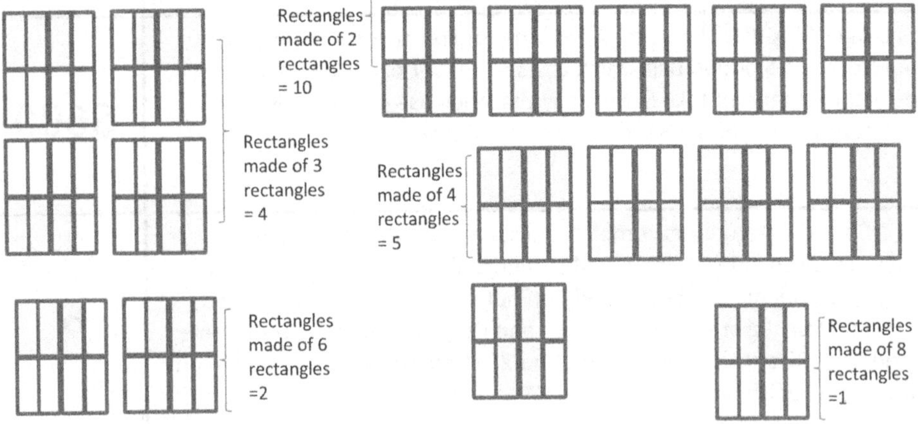

PERENNIAL MATH

TEST #3

January 2013 Advanced Level Solutions

4) Joe can mow his mother's lawn in 60 minutes. Ricky can mow it in 90 minutes.

$$\text{Time} = \frac{1}{\frac{1}{X}+\frac{1}{Y}} \qquad \text{Time} = \frac{1}{\frac{1}{60}+\frac{1}{90}} \qquad \frac{1}{5/180}$$ = **36 minutes**

36

5) There are 20 potatoes in a 5 lb. bag, which would yield 10 orders of fries. There are 400 lbs. of potatoes, so that would take 80 of the 5lb. Bags. 80 x 10 orders per bag is 800 orders. 800 x $1 is $800. However, the bags of potatoes cost 80 x $2.50 = $200. So the restaurant would make $800-$200 = **$600** for the day.

$600

6) All regions total 50 students. Label the totals for each category:30 students liked Cats, 32 liked Dogs, and 28 liked Fish. Fill in these areas: 4 students liked Cats and Fish but not Dogs. 12 students liked Dogs and Fish but not Cats. Use these to solve for missing areas:10 students liked Cats and Fish [Region V + 4]. 18 students liked Cats and Dogs [Region V + 12]. Once you solve for Region V then you can solve for the fish to get a total of 28. See diagram below.

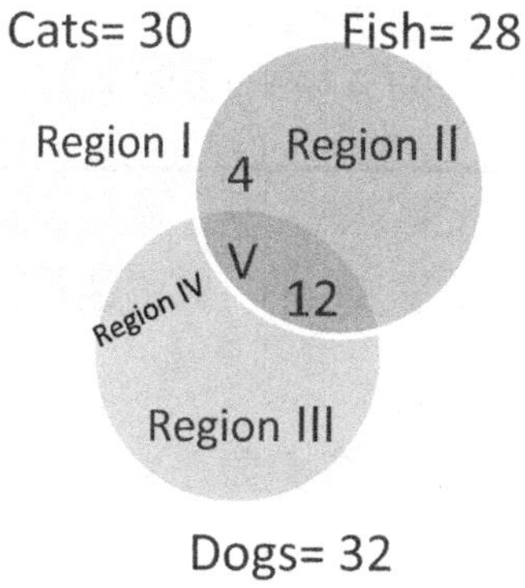

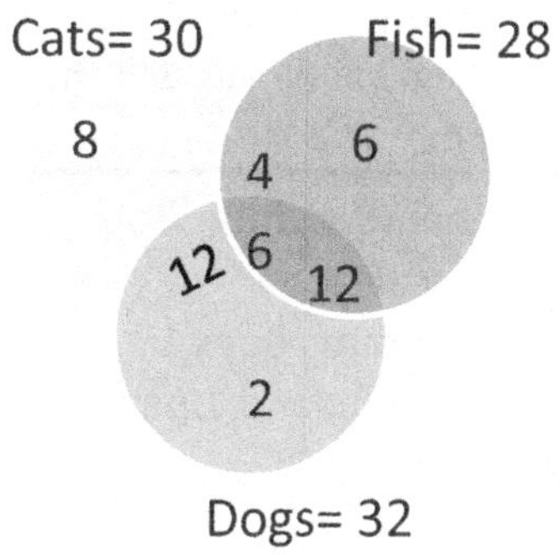

6

PERENNIAL MATH

TEST #4

February 2013 Advanced Level

1) The first stage of a rocket burns 32 seconds longer than the second stage. If the total burning time for both stages is 148 seconds, how many seconds does the first stage burn?	
2) Krishna has $11.25, all dimes and quarters, in her piggy bank. She has 32 more dimes than quarters. How many dimes does she have?	
3) At his party, Tariq had 1087 jellybeans in his jar. For every 12 colorful jellybeans that he gave to friends, there was 1 licorice that he would throw away. He still had 34 jellybeans left in his jar after the party. How many jellybeans did he give away?	
4) At the hotdog stand there are 5 choices of toppings: ketchup, mustard, relish, onions, and chili. You can select as many toppings as you want. How many possible topping combinations are there?	
5) Students were building a tower from blocks. The top level is a 1 x 2 blocks rectangle. The level underneath is a 2 x 3 blocks rectangle. The next level underneath is a 3 x 4 blocks rectangle. How many total blocks will they need for a tower that is 12 levels high?	
6) Use the information from the three scales. How many grams does 1 orange, 1 apple, and 1 pear weigh? 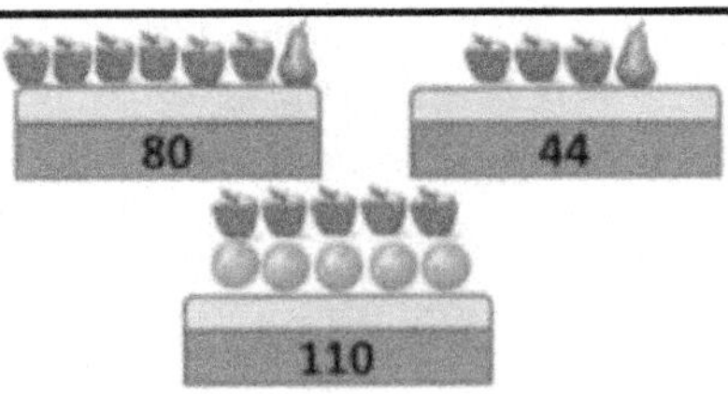	

PERENNIAL MATH TEST #4

February 2013 Advanced Level Answers

1) The first stage of a rocket burns 32 seconds longer than the second stage. If the total burning time for both stages is 148 seconds, how many seconds does the first stage burn?	**90**
2) Krishna has $11.25, all dimes and quarters, in her piggy bank. She has 32 more dimes than quarters. How many dimes does she have?	**55**
3) At his party, Tariq had 1087 jellybeans in his jar. For every 12 colorful jellybeans that he gave to friends, there was 1 licorice that he would throw away. He still had 34 jellybeans left in his jar after the party. How many jellybeans did he give away?	**972**
4) At the hotdog stand there are 5 choices of toppings: ketchup, mustard, relish, onions, and chili. You can select as many toppings as you want. How many possible topping combinations are there?	**31**
5) Students were building a tower from blocks. The top level is a 1 x 2 blocks rectangle. The level underneath is a 2 x 3 blocks rectangle. The next level underneath is a 3 x 4 blocks rectangle. How many total blocks will they need for a tower that is 12 levels high?	**728**
6) Use the information from the three scales. How many grams does 1 orange, 1 apple, and 1 pear weigh? 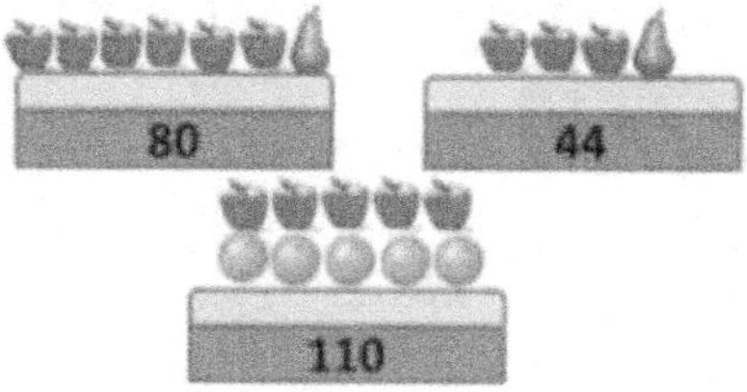	**30**

PERENNIAL MATH

TEST #4

February 2013 Advanced Level Solutions

1)Let the first stage be x + 32 and the second stage be x.

Then (x + 32) + x = 148.

2x + 32 – 32 = 148 – 32. 2x = 116. 2x/2 = 116/2. X = 58.

The first stage is x + 32, so 58 + 32 = **90 seconds.**

90

2) Krishna has $11.25.She has 32 more dimes than quarters.

0.25q+0.10(q+32)=11.25. We can then solve for q, and then for q+32.

0.25q+0.10(q+32)=11.25.

.35q + 3.20 = 11.25

.35q + 3.20 -3.20 = 11.25 – 3.20

.35q= 8.05

.35q/.35 = 8.05/.35

q = 23.

Since there are 32 more dimes than quarters, there are 23 + 32= **55**

55

3) At his party, Tariq had 1087 jellybeans in his jar. For every 12 colorful jellybeans that he gave to friends, there was 1 licorice that he would throw away. He still had 34 jellybeans left in his jar after the party. How many jellybeans did he give away? Take the total and subtract how many were left over. Divide this by thirteen because he gave away 12 and threw 1 away which is 13 portions.

$$N = \frac{1087-34}{13}$$

N= 81 Multiply by 12 because he gave away 12 times as many as he threw away. ***He gave away 972***.

972

PERENNIAL MATH

TEST #4

February 2013 Advanced Level Solutions

4) At the hotdog stand there are 5 choices of toppings: ketchup, mustard, relish, onions, and chili. You can select as many toppings as you want. How many possible topping combinations are there?
With five toppings there are thirty-one combinations
Ketchup=K, Mustard=M, Relish=R, Onions=O, Chili=C
K, KM, KR, KO,KC, KMR, KMO, KMC, KRO, KRC, KOC, KMRO, KMRC, KMOC,KROC, KMRCO
M, MR, MO, MC, MRO, MRC, MOC, MROC,
R, RO, RC, ROC
O, OC
C

31

5) The top level is a 1 x 2 blocks rectangle. The level underneath is a 2 x 3 blocks rectangle. The next level underneath is a 3 x 4 blocks rectangle. How many total blocks will they need for a tower that is 12 levels high?

Level 1	1x2	2
Level 2	2x3	6
Level 3	3x4	12
Level 4	4x5	20
Level 5	5x6	30
Level 6	6x7	42
Level 7	7x8	56
Level 8	8x9	72
Level 9	9x10	90
Level 10	10x11	110
Level 11	11x12	132
Level 12	12x13	156
	Total	728

728

6) (3A + P) = 44
(3A + P) + 3A = 80
80 – 44 = 36, so 3A =36, A = 12
If A = 12, then 5A + 5O = 110
5(12) + 5O = 110, 110- 60 = 50, so O = 10
Substitute 12 in for A to solve for P
If 3A + P = 44, then 3(12) + P = 44, 44-36 = 8 So P = 8
O + A + P= 10 + 12 + 8 = **30**

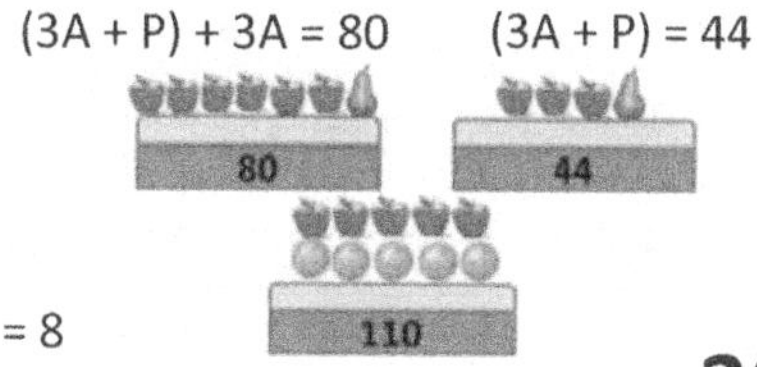

30

PERENNIAL MATH

TEST # 1

November 2013 Advanced Level

1) Two prime numbers y and z have a sum of 38. If $y > z$, find the value of $y - z$.	
2) Alex, Holly, Nigel, Carrie, Samantha and Trent went bowling. Their average score for the first game was 142. As they were finishing, Shirley joined them and scored 121. What is the average for the group now?	
3) If two sides of a square field were increased by four feet, as shown in the diagram, the area of the field would increas by 136 square feet. How many square feet a in the area of the original field (shaded squa 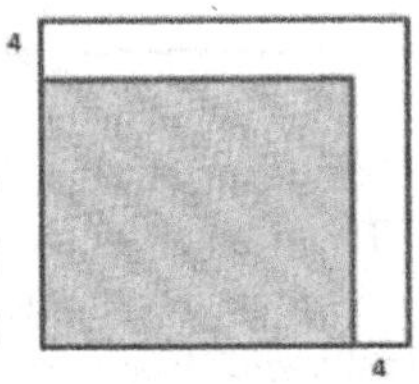	
4) Beverly is working on her class schedule. If order does not matter, how many ways can Beverly choose three electives from the five choices (art appreciation, video production, computer graphics, music theory, and home economics)?	
5) A rectangle whose length is twice its width has an area of 1800 cm^2. If the length is increased by ten percent and the width is decreased by ten percent, how many square centimeters are in the area of the new rectangle?	
6) Consecutive numbers are whole numbers that follow in order such as 3, 4, 5. Find the smallest of the five consecutive numbers whose sum is 200.	

PERENNIAL MATH TEST #1

November 2013 Advanced Level Answers

1) Two prime numbers *y* and *z* have a sum of 38. If $y > z$, find the value of $y - z$.	**24**
2) Alex, Holly, Nigel, Carrie, Samantha and Trent went bowling. Their average score for the first game was 142. As they were finishing, Shirley joined them and scored 121. What is the average for the group now?	**139**
3) If two sides of a square field were increased by four feet, as shown in the diagram, the area of the field would increas by 136 square feet. How many square feet a in the area of the original field (shaded squa 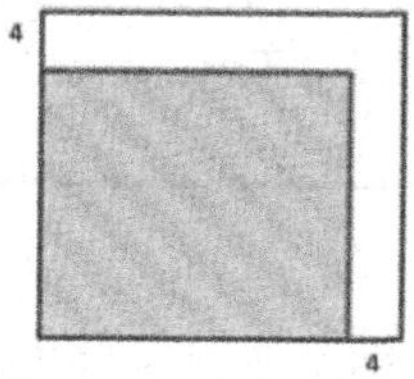	**225**
4) Beverly is working on her class schedule. If order does not matter, how many ways can Beverly choose three electives from the five choices (art appreciation, video production, computer graphics, music theory, and home economics)?	**10**
5) A rectangle whose length is twice its width has an area of 1800 cm^2. If the length is increased by ten percent and the width is decreased by ten percent, how many square centimeters are in the area of the new rectangle?	**1782**
6) Consecutive numbers are whole numbers that follow in order such as 3, 4, 5. Find the smallest of the five consecutive numbers whose sum is 200.	**38**

PERENNIAL MATH TEST # 1

November 2013 Advanced Level Solutions

1) The number 1 is not a prime number. Therefore, 37 and 1 cannot be considered. The two primes must be 31 and 7. The difference $y - z$ is 31 – 7 = **24**.

24

2) Using the formula Average = Total sum of scores/number of scores, their average is 142 = x/6. The total number of points they scored together was 6 x 142= 852 total points. Adding Shirley's score of 121 to 852 = 973 total points. Calculating the new average then becomes 987 ÷ 7 or **139.**

139

3) The corner square is 4 by 4 or sixteen square feet. The rest of the new or added space is 120 square feet. Each rectangle is 60 sq. ft. with dimensions 4 by 15. This gives the square a side of 15 feet. With a 15 by 15 square we have total
area **225 ft²** in the original shaded square.

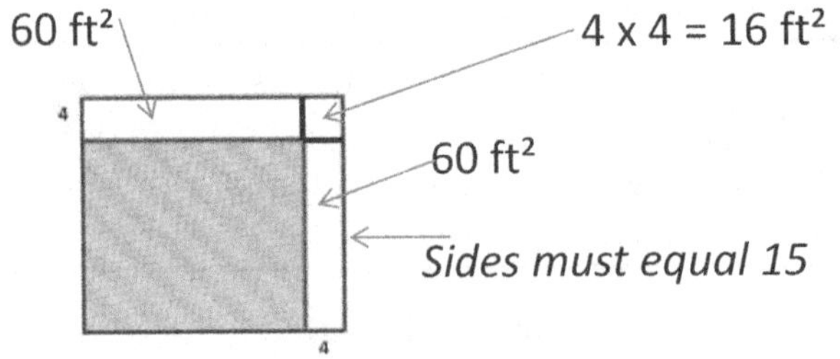

225

PERENNIAL MATH

TEST # 1

November 2013 Advanced Level Solutions

4) Label the classes A, B, C, D and E. Make an organized list to eliminate errors of omitting answers or repeating answers. Combinations are ABC, ABD, ABE, ACD, ACE, ADE, BCD, BCE, BDE or CDE. So there are **10** ways to choose 3 electives from the 5 classes.

A	B	C	D	E
ABC	BCD	CDE		
ABD	BCE			
ABE	BDE			
ACD				
ACE				
ADE				

10

5) A rectangle whose length is twice its width and has an area of 1800 cm^2 has dimensions of 60 cm by 30 cm. If the length is increased by 10% then 10% of 60 is 6, so the new length is 66. If the width is decreased by 10% then 10% of 30 is 3, so the new width is 27. The area of the new rectangle would be 66 cm x 27 cm = **1782 cm^2**

1782

6) To find the smallest of the five consecutive numbers whose sum is 200, divide 200 by 5. This will give you the mean or average which in this case is the middle number. Since 40 is the mean, the five numbers are 38, 39, 40, 41, 42. Therefore, the smallest number is **38.**

38

PERENNIAL MATH

TEST #2

December 2013 Advanced Level

1) $8 \times 7 - 4^2 + 9 \times 2 =$

2) The ratio of apples to oranges in a box was 4 to 9. If there were 48 apples and one third of the oranges were rotten, how many oranges were not rotten?

3) An arrow is formed in this 6 x 6 cm square by joining the bottom corners to the midpoint of the top edge and the center of the square. How many square centimeters are in the area of the arrow?

4) Connie is three times as old as Vickie is now. In 30 years Connie will be 18 years older than Vickie will be then. What is Connie's age now?

5) A palindrome is a number that reads the same forwards and backwards such as 1441 or 32523. Find the average of the fifteen smallest four digit palindromes.

6) Find the number of addition problems in which a two digit number is added to a second two digit number to yield a two digit answer that is less than 50. *(10 + 15 and 15 + 10 are considered as different problems.)*

PERENNIAL MATH

TEST #2

December 2013 Advanced Level Answers

1) $8 \times 7 - 4^2 + 9 \times 2 =$	**58**
2) The ratio of apples to oranges in a box was 4 to 9. If there were 48 apples and one third of the oranges were rotten, how many oranges were not rotten?	**72**
3) An arrow is formed in this 6 x 6 cm square by joining the bottom corners to the midpoint of the top edge and the center of the square. How many square centimeters are in the area of the arrow? 	**9**
4) Connie is three times as old as Vickie is now. In 30 years Connie will be 18 years older than Vickie will be then. What is Connie's age now?	**27**
5) A palindrome is a number that reads the same forwards and backwards such as 1441 or 32523. Find the average of the fifteen smallest four digit palindromes.	**1,738**
6) Find the number of addition problems in which a two digit number is added to a second two digit number to yield a two digit answer that is less than 50. *(10 + 15 and 15 + 10 are considered as different problems.)*	**465**

PERENNIAL MATH

TEST # 2

December 2013 Advanced Level Solutions

1) Use the order of operations *PEMDAS*

$8 \times 7 - 4^2 + 9 \times 2 =$

56 – 16 + 18 = ***58***

58

2) The ratio of apples to oranges is 4:9. If there are 48 apples, then there must be (9 x 12) or 108 oranges. If 1/3 of the oranges are rotten then 108/3 = 36. Thirty-six of the oranges are rotten and 108 – 36 = **72 oranges** that are good.

72

3) Since the sides of the square are 6 centimeters long, the area of the square is 36 cm^2. The area of the large triangle is 18 cm^2 because it is half of the square. The area of the small triangle is 9 cm^2 because it is a quarter of the square.

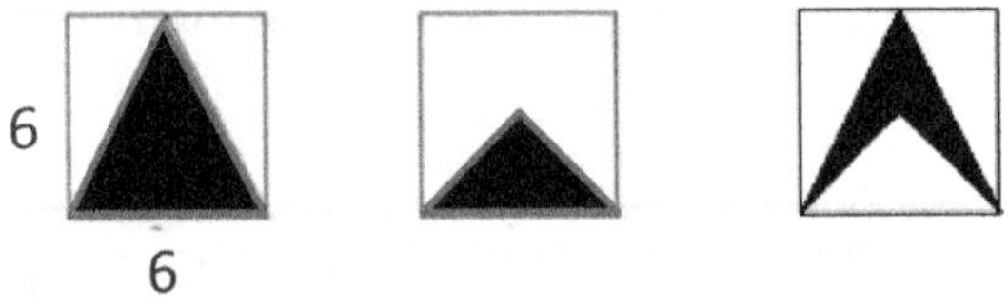

Hence the area of the arrow is 18 cm^2 - 9 cm^2 = **9 cm^2**.

9 cm^2

PERENNIAL MATH

TEST # 2

December 2013 Advanced Level Solutions

4) Make a table to compare the current ages of the girls and their ages in 30 years. **Connie is 27** and Vickie is 9 right now. In 30 years Connie will be 18 years older than Vickie (57 – 39 = 18).

Connie	Vickie	Connie + 30	Vickie + 30	C - V
3	1	33	31	2
6	2	36	32	4
9	3	39	33	6
12	4	42	34	8
15	5	45	35	10
18	6	48	36	12
21	7	51	37	14
24	8	54	38	16
27	9	57	39	18

27

5) Add the first 15 four digit palindromes : 1001, 1111, 1221, 1331, 1441, 1551, 1661, 1771, 1881, 1991, 2002, 2112, 2222, 2332, 2442 = 26,070

To find the average divide 26,070 ÷15 = **1,738**

1,738

6) If you list the two digit numbers and the numbers that can be added to them and make a chart you will notice that there are 30 possible sums starting with the number 10 and then the number of sums goes down by 1 all the way down to the number 39. There are **465 possible sums** of two digit numbers whose sum is less than 50.

Two-digit number																															number of sums
10	10	11	12	13	14	15	16	17	18	19	20	21	22	23	24	25	26	27	28	29	30	31	32	33	34	35	36	37	38	39	30
11	10	11	12	13	14	15	16	17	18	19	20	21	22	23	24	25	26	27	28	29	30	31	32	33	34	35	36	37	38		29
12	10	11	12	13	14	15	16	17	18	19	20	21	22	23	24	25	26	27	28	29	30	31	32	33	34	35	36	37			28
13	10	11	12	13	14	15	16	17	18	19	20	21	22	23	24	25	26	27	28	29	30	31	32	33	34	35	36				27
14	10	11	12	13	14	15	16	17	18	19	20	21	22	23	24	25	26	27	28	29	30	31	32	33	34	35					26
15	10	11	12	13	14	15	16	17	18	19	20	21	22	23	24	25	26	27	28	29	30	31	32	33	34						25
16	10	11	12	13	14	15	16	17	18	19	20	21	22	23	24	25	26	27	28	29	30	31	32	33							24
17	10	11	12	13	14	15	16	17	18	19	20	21	22	23	24	25	26	27	28	29	30	31	32								23
18	10	11	12	13	14	15	16	17	18	19	20	21	22	23	24	25	26	27	28	29	30	31									22
19	10	11	12	13	14	15	16	17	18	19	20	21	22	23	24	25	26	27	28	29	30										21
20	10	11	12	13	14	15	16	17	18	19	20	21	22	23	24	25	26	27	28	29											20
21	10	11	12	13	14	15	16	17	18	19	20	21	22	23	24	25	26	27	28												19
22	10	11	12	13	14	15	16	17	18	19	20	21	22	23	24	25	26	27													18
23	10	11	12	13	14	15	16	17	18	19	20	21	22	23	24	25	26														17
24	10	11	12	13	14	15	16	17	18	19	20	21	22	23	24	25															16
25	10	11	12	13	14	15	16	17	18	19	20	21	22	23	24																15
26	10	11	12	13	14	15	16	17	18	19	20	21	22	23																	14
27	10	11	12	13	14	15	16	17	18	19	20	21	22																		13
28	10	11	12	13	14	15	16	17	18	19	20	21																			12
29	10	11	12	13	14	15	16	17	18	19	20																				11
30	10	11	12	13	14	15	16	17	18	19																					10
31	10	11	12	13	14	15	16	17	18																						9
32	10	11	12	13	14	15	16	17																							8
33	10	11	12	13	14	15	16																								7
34	10	11	12	13	14	15																									6
35	10	11	12	13	14																										5
36	10	11	12	13																											4
37	10	11	12																												3
38	10	11																													2
39	10																														1
																														Total Sums	465

465

PERENNIAL MATH

TEST #3

January 2014 Advanced Level

1) What is the tens digit of the sum $1^2 + 2^2 + 3^2 + ... + 10^2$?

2) John was producing collections of numbers. He wanted no more than six numbers in a collection and needed all the numbers in each collection to sum to 20. The numbers in each group could only be characterized as positive odd integers. How many possible collections could he form?
Note that 1 + 1 + 3 + 3 + 3 + 9 and 9 + 1 + 3 + 1 + 3 + 3 are considered to be the same collection.

3) Complete this magic square so that all rows, columns, and diagonals have the same sum. What number will N represent?

	55	
25	95	N

4) The ratio of Kyle's marbles to Brad's marbles was 4:1. Kyle gave half his marbles to Brad. Then Kyle discovered that he had 12 fewer marbles than Brad. How many marbles did Kyle have to start with?

5) A girl's father paid for one-third of her prom dress that originally cost $600. The dress was purchased on sale at 20% off. How much did the father pay?

6) Three pickers can harvest 30 gallons of peppers in four hours. At this rate how long will it take five pickers to harvest 75 gallons of peppers?

PERENNIAL MATH

TEST #3

January 2014 Advanced Level Answers

1) What is the tens digit of the sum $1^2 + 2^2 + 3^2 + \ldots + 10^2$?

8

2) John was producing collections of numbers. He wanted no more than six numbers in a collection and needed all the numbers in each collection to sum to 20. The numbers in each group could only be characterized as positive odd integers. How many possible collections could he form?
Note that 1 + 1 + 3 + 3 + 3 + 9 and 9 + 1 + 3 + 1 + 3 + 3 are considered to be the same collection.

30

3) Complete this magic square so that all rows, columns, and diagonals have the same sum. What number will N represent?

	55	
25	95	N

45

4) The ratio of Kyle's marbles to Brad's marbles was 4:1. Kyle gave half his marbles to Brad. Then Kyle discovered that he had 12 fewer marbles than Brad. How many marbles did Kyle have to start with?

48

5) A girl's father paid for one-third of her prom dress that originally cost $600. The dress was purchased on sale at 20% off. How much did the father pay?

160

6) Three pickers can harvest 30 gallons of peppers in four hours. At this rate how long will it take five pickers to harvest 75 gallons of peppers?

6 hrs

PERENNIAL MATH TEST #3

January 2014 Advanced Level Solutions

1) 1 + 4 + 9 + 16 + 25 + 36 + 49 + 64 + 81 + 100 = 385.
Therefore, the tens digit is **8**.

8

2) Make an ordered list.

19+1
17+3
17+1+1+1
15+5
15+3+1+1
15+1+1+1+1+1
13 + 7
13+5+1+1
13+3+3+1
13+3+1+1+1+1
11+9
11+7+1+1
11+5+3+1
11 + 5 + 1+ 1 + 1 + 1
11+3+3+3
11+3+3+1+1+1

9 + 9 + 1 + 1
9 + 7 + 3 + 1
9 + 7 + 1 + 1 + 1 + 1
9 + 5 + 5 + 1
9 + 5 + 3 + 3
9 + 5 + 3 + 1 + 1 + 1
9 + 3 + 3 + 3 + 1 + 1
7 + 7 + 5 + 1
7 + 7 + 3 + 3
7 + 5 + 5 + 3
7 + 5 + 5 + 1 + 1 + 1
5 + 5 + 5 + 5
5 + 5 + 5 + 3 + 1 + 1
5 + 5 + 3 + 3 + 3 + 1
5 + 3 + 3 + 3 + 3 + 3

There are **31 sums** of six or less positive odd integers.

31

3) Since N is shared by 25 + 95 + N and N + 55 + X, then 25 + 95 = 55 + X so X = 65. Then you can solve for Z.
65 + Y + Z = 95 + 55 + Y, then 65 + Z = 95 + 55. Z = 85. Now we know what all sides will equal because we have solved for the diagonal 25 + 55 + 85 = 165. Now we can solve for N by subtracting. 165 – (25 + 95) = N. **N= 45**

X	Y	Z
	55	
25	95	N

65	15	85
75	55	35
25	95	45

45

PERENNIAL MATH TEST #3

January 2014 Advanced Level Solutions

4)

The ratio of Kyle's marbles to Brad's marbles was 4:1. When Kyle gave half his marbles to Brad the ratio became 2:3 and Kyle discovered that he had 12 fewer marbles than Brad. One portion is equal to 12 marbles. So Kyle had 4 portions to begin which was **48 marbles**.

48

5) Since the dress was 20% off, the daughter/father paid 80% (100-20) of \$600. Which is calculated by \$600*0.80 = \$480
Since the father paid for one-third of the dress, he paid **\$160** (480/3)

160

6) Three pickers harvest 30 gallons in four hours or each picker harvests 10 gallons in four hours. Ten gallons in four hours converts to 2.5 gallons each hour per picker. At this rate, five pickers will harvest (5 x 2.5) = 12.5 gallons each hour. 75 ÷ 12.5 = **6, or six hours** are needed by five pickers to harvest 75 gallons of peppers.

ALTERNATE SOLUTION

In the first scenario, 3 pickers harvest for 4 hours a piece, which totals to 12 picker hours to produce 30 gallons. Set up a proportion relating picker hours to gallons produced.

$$\frac{12}{30} = \frac{x}{75}$$

Cross multiply and solve
x= 30 (30 picker hours produces 75 gallons)

Since there were 5 pickers, then each picker worked **6 hours** (30 ÷ 5)

6 hours

PERENNIAL MATH TEST #4

February 2014 Advanced Level

1) How many rectangles can be made that have an area of 144 square units and the length and widths are natural numbers?	
2) What is the difference between the largest three-digit prime number and the largest three-digit number that only has three positive factors?	
3) Trevor played a game in which he could score either 9 or 14 points on each turn. If Trevor scored exactly 100 points, on how many turns did he score 9 points?	
4) The soccer field is sufficiently lit when at least two of the six sets of lights that surround the field are on. In how many different ways can lights be turned on so that the field is sufficiently lit?	
5) Austin's weekly paycheck is 15% less than Eric's paycheck. Their paychecks have a combined total of $777. How much is Austin's paycheck worth?	
6) If you continue this pattern, what would be the sum of the circles in the 15th and 21st items in the sequence? 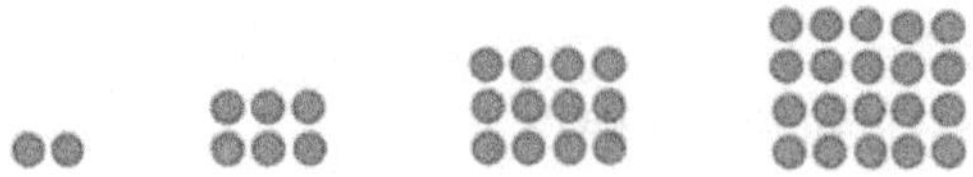	

PERENNIAL MATH

TEST #4

February 2014 Advanced Level Answers

1) How many rectangles can be made that have an area of 144 square units and the length and widths are natural numbers?	**8**
2) What is the difference between the largest three digit-prime number and the largest three-digit number that only has three positive factors?	**36**
3) Trevor played a game in which he could score either 9 or 14 points on each turn. If Trevor scored exactly 100 points, on how many turns did he score 9 points?	**8**
4) The soccer field is sufficiently lit when at least two of the six sets of lights that surround the field are on. In how many different ways can lights be turned on so that the field is sufficiently lit?	**57**
5) Austin's weekly paycheck is 15% less than Eric's paycheck. Their paychecks have a combined total of \$777. How much is Austin's paycheck worth?	**\$357**
6) If you continue this pattern, what would be the sum of the circles in the 15th and 21st items in the sequence? 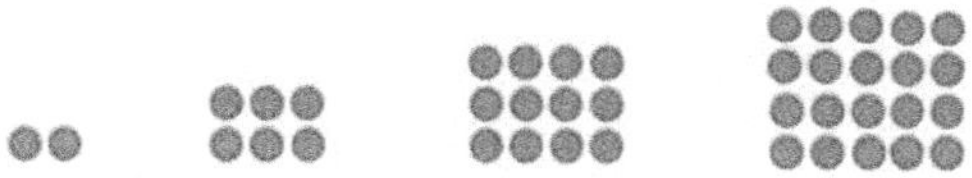	**702**

PERENNIAL MATH

TEST # 4

February 2014 Advanced Level Solutions

1) Make a list of the possible lengths and widths of rectangles with an area of 144 units^2: (1, 144), (2, 72), (3, 48), (4, 36), (6, 24), (8, 18), (9, 16), (12, 12). There are **8 different rectangles** whose dimensions equal 144 square units. Don't forget that a square is also a rectangle.

8

2) The largest three digit prime number is 997. The largest three digit number with three positive factors is 961. (The factors of 961 are 1, 31, 961 In order for a number to have three factors, it would have to also be a square made from a prime number and 31 is the highest prime to give a three digit product). The difference between the largest prime number and the largest three digit number with three factors is 997-961 = **36**.

36

3) In order for the point total to be 100, there must be an even number of 9 points scored. Make a quick chart to see what the answer is:

9 points	14 points	Total Score
2	6	102
4	5	106
6	3	96
8	2	100

8

Trevor scored 9 points on **8 turns**.

PERENNIAL MATH TEST #4

February 2014 Advanced Level Solutions

4) The problem states that at least 2 lights must be lit so we can have combinations of 2 lights, 3 lights, 4 lights, 5 lights or even all 6 lights on. Label the lights A through F. Here are the possible combinations. There are 15 (2-light combinations) + 20 (3-light combinations) + 15 (4-light combinations + 6 (5-light combinations) + 1 (6-light combination) for **57 total combinations** that have at least 2 lights.

2 lights	3 lights	4 lights	5 lights	6 lights
AB	ABC	ABCD	ABCDE	ABCDEF
AC	ABD	ABCE	ABCDF	1
AD	ABE	ABCF	ABCEF	
AE	ABF	ABDE	ABDEF	
AF	ACD	ABDF	ACDEF	
BC	ACE	ABEF	BCDEF	
BD	ACF	ACDE	6	
BE	ADE	ACDF		
BF	ADF	ACEF		
CD	AEF	ADEF		
CE	BCD	BCDE		
CF	BCE	BCDF		
DE	BCF	BCEF		
DF	BDE	BDEF		
EF	BDF	CDEF		
15	BEF	15		
	CDE			
	CDF			
	CEF			
	DEF			
	20			

57

5) X+(1-.15)X = 777

X+.85X = 777

1.85X = 777

X =777/1.85

X = $420 Eric's paycheck.

$777 - $420 = **$357 Austin's paycheck**

$357

6)

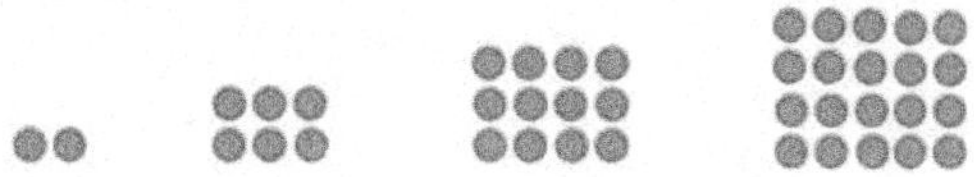

The characters in the sequence given are 1 x 2, 2 x 3, 3 x 4, and 4 x 5. The characters in the sequence follow the rule: n x (n + 1). The 15th character would be 15 x 16 = 240 circles. The 21st character would have 21 x 22 = 462 circles. Their sum would have 240 + 462 = **702 circles**.

702